EARNED VALUE MANAGEMENT BASICS

A LAYMAN'S GUIDE TO EARNED VALUE MANAGEMENT

Phill C. Akinwale

Earned Value Management Basics
Published by Praizion Media
P.O. Box 22241, Mesa, AZ 85277
E-mail: info@praizion.com
www.praizion.com

Author:
Phillip Akinwale, MSc, PMP, PMI-RMP, PMI-SP, CAPM, CSM

PMI®, PMBOK®, CAPM®, PMP® and PgMP® are trademarks and certification marks of the Project Management Institute, which are registered in the United States and other nations.

Scenarios, characters, institutions and/or entities used in this book are fictitious. Any similarity, likeness or resemblance to persons living or dead, actual events, institutions and/or entities, is purely coincidental and unintentional.

The author and publisher accept no liability, losses or damages of any kind caused or alleged to be caused directly or indirectly by this publication.

Printed in the United States of America

Table of Contents

Introduction

E arned Value Management is one of those areas of project management that some project managers often dread on professional exams largely because they have never used it on real-world projects. Some are intimidated at the sheer mention of the acronym "EVM". I was one of such project managers who knew EVM on paper but had never used it until one day in 2005, I received a call from a recruiter asking me if I would be interested in working for a major Aerospace company. Apart from the interview which grilled me for all I was worth on knowledge of the subject (which I had great working knowledge of), within weeks, I was fully immersed into a world of Earned Value Management on over 40 projects in a huge program of projects. Earned Value Management metrics became my breakfast, lunch and dinner daily. It was such a huge part of my work and my first consulting stint at a major Aerospace company that it became a way of life and second nature. I gained a deep understanding of all the formulas, metrics and approaches. I still recall the legacy application with the blue screen which we used for WBS creation, EVM and project controls. My colleague on the engineering team introduced me to the arcane arts of Excel macros, using them to display high-tech fancy graphs and charts that would hold the most stoic managers spellbound.

It was the excitement and first-hand experience in seeing what Earned Value Management could do that inspired me to write a book on it with fictional project management team characters to demonstrate its value and help project managers who are new to the subject understand it better. I hope you enjoy it!

GO EVM'ERs!

"For the enlightened project manager, Earned Value Management is more than a methodology. It is a way of life."

Phill C. Akinwale

Chapter 1: What is Earned Value?

You don't know EVM and you're our PM?

"What is Earned Value Management?" Brea asked me looking extremely perplexed. She had just started working with the team and was absolutely clueless about the topic. Being a Masters student, I joked that it was certainly not half as tough as some of her MBA courses. She was not convinced. "How do you arrive at so many different values with just 3 values?" She inquired.

My mind travelled back to 2005 when I had just taken the PMP exam, and the term earned value seemed to be very esoteric and "out there" and very hard to understand. I knew how to solve problems on the exam, but what I didn't really know was how to implement an Earned Value Management System and what it really entailed to manage programs and projects with earned value. It wasn't until I had the opportunity of working for a major Aerospace firm shortly after becoming PMP® Certified in January 2006 that I began to understand what exactly earned value could be used for and its main advantage being the number one approach for measuring project performance. In fact, if I were to manage any project above a certain threshold, my number one priority for measuring project fulfillment would be EVM because it is absolutely phenomenal. It cuts across time, cost, percent complete, and when handled by a good project manager looking straight down the project with eagle-eyed vision, there is so much that one can glean about a project's performance, either good or bad.

"So how do you do it?" Brea prodded interrupting me from my retrospective journey down memory lane. "Why do we use this to manage our projects in the first place?" she inquired. "Why not simply budget spent versus budget planned? And how did earned value management become so important?"

"Well, Brea" I said opening up the EVMS (Earned Value Management System), "Earned value management originated back in the 1960s. It first of all emerged as a financial analysis specialty in the U.S. government programs back at that time, but ever since the 60s, earned value management has evolved to become a significant part of project management and engineering, especially."

"So this has been around since the 60s?" She asked in amazement.

"Yes. In fact, most engineering projects that are high visibility government projects run by various agencies are managed using EVM. Earned Value Management is extremely important to today's project manager working for or with government agencies."

"Why?" She asked curiously.

"Because the United States Office of Management and Budget (OMB) have very recently mandated the use of EVM across all government agencies and we're talking about, not just contractors that work for these agencies, but also internal projects for these agencies."

By this time Brea had at least one full page of notes and was scribbling away furiously like a student learning from a professor.

I continued, "So it becomes even more important for project managers or contractors working for or with the U.S. government to understand what EVM is."

"Maybe when I take my PMP® exam, I will learn more about it but I need something I can learn fast to do my job." She said with desperation.

"Now, Brea, an overview of Earned Value Management was included in the first *PMBOK® Guide*, and in subsequent editions, Earned Value Management has been highlighted. Earned Value Management appears on the PMP® exam. So it's very important to understand its implication to projects, and if you needed to use EVM at any point in time (like you do now), at least you will have that base knowledge to help you like I did from studying for the PMP® Exam." At this point, Brea proceeded to flip open her laptop and began typing furiously.

"Is there a book you can recommend to better master this? I would like to use it on a small personal project I am working on."

"My book on EVM will be out soon." I said. She laughed dismissively. Little did she know that I already had the draft copy of some of my EVM musings I was editing. It would be a few years later that I finalized and published the book. On a nearby post-it note, I wrote down the web address for the Defense Acquisition University (DAU) which I had thoroughly combed through many months before. It proved to be an invaluable repository of knowledge on EVM.

"The good thing about EVM," I continued, "Is that it can be scaled up or down. Depending on the size of project you're working on and its complexity, you could introduce EVM at various levels."

"Great! I will get on that site right away and begin learning."Brea said.

"Before you do that, let me show you a couple of personal notes I have written on the topic." I slid the first few pages of my draft manuscript on EVM out of my folder. "Here you go!" I said handing it to her. She literally snatched the pages out of my hands in eagerness. The pages read:

Earned Value Definitions

The term "earned value" is widely used to describe a methodology for project performance tracking known formally as Earned Value Management. Earned Value Management involves using schedule, cost and scope information to define project status and likely future outcomes such as:

- Is a project on schedule?
- Is a project on budget?
- How much is a project team likely to end up spending on a project based on current project performance?

Earned Value Management enables you estimate the value of the work accomplished and indicates the project team's efficiency. It demarcates budget as "earned" only when a specific activity, event, portion of the work or milestone is complete.

Vital Point - Earned Value calculations employ four basic metrics:

- Planned Value (PV), also described as budgeted cost of work scheduled (BCWS)

- Earned Value (EV), also described as budgeted cost of work performed (BCWP)

- Actual Cost (AC), also described as actual cost of work performed (ACWP)

- Budget at completion (BAC)

These metrics are based on time and do change over time on most projects. When calculating earned value metrics, you MUST use values from the same time period. For example values for AC, EV and PV must all be from January to June. You cannot combine values arbitrarily such as an EV from

14

December to May, a PV from February to July and an AC from May to November. Accuracy of earned value depends on the accuracy of reporting the percentage of work done.

"Wow!" she exclaimed. "This is great. I feel I understand a lot already!" she said beaming. "You should write a book on this!" She obviously took what I said earlier as a joke. Little did she know she was reading the first few pages of what would become Earned Value Management Basics!

Since then, Brea has become a PMP® and has moved on to bigger things having become very proficient in Earned Value and handed over her job to someone else.

So, now, roll up your sleeves and get ready to learn a little bit more about EVM and how to implement it on your projects.

Team Discussion

1. What is EVM?

2. EVM uses _____, cost and _____ information to predict _____

3. Discuss 4 major EVM metrics mentioned in this chapter.

4. Budgeted Cost of Work Scheduled is also known as?

5. Budgeted Cost of Work Performed is also known as?

6. AC is also known as?

7. EVM first emerged in what decade?

8. One good thing about EVM is that it is _____ and can be used on both large and small projects.

"Reduce your plan to writing. The moment you complete this, you will have definitely given concrete form to the intangible desire."

Napoleon Hill

Chapter 2: EVM Philosophy 101

Gee! The PMO has no clue about the WBS? I need a raise!

On Day 1 of our first Earned Value Management class, Brea turned up with a couple of other project managers from our immediate business unit and four more PMs from IT! I was impressed that IT folks in their own little world of Agile driven project management were willing to learn more about EVM!

It was only an hour-long session but I was determined to get as much mileage as possible out of the class since projects had literally consumed us.

There were WBS charts to be completed for the next fiscal year program of projects and lots of other things to do. Training was the last thing on most employees' minds but this subset of the team had decided they needed to know more about it to do their jobs.

"Earned Value Management", I started off, "Is all about managing well defined projects with concrete proof of performance and tangible time-phased targets. I pointed to the whiteboard which had the following written out:

The underlying philosophy of Earned Value Management is that: You should get the full value of what you pay for but if you do not get full value of what you paid for, then analyze the following:

I.   What is the true value of work accomplished on the project within a particular time period?

II.   How much was paid for the work?

III.   Is there a variance between I and II?

IV.   What was the scheduled duration for work accomplished?

V. What was the actual duration for the work accomplished?

VI. Is there a variance between IV and V?

VII. If so, why did this happen? Why did you not obtain the true value of what you paid for?

VIII. Realize that your team cannot keep losing time and money.

IX. Decide what you will do to correct the causes of these variances on current and future projects.

X. TAKE CORRECTIVE OR PREVENTIVE ACTION!

"Ok I get that." Said Kirk from the IT department. But do you have some practical examples? "Sure" I replied. I wrote out the following example on the whiteboard.

Example:

a. A Project Manager hires a painter at $5 per hour to paint 1 wall.

b. The painter estimates the wall would be painted in 2 hours but agrees to be paid hourly.

c. The painter begins the project but it rains and the painter can do nothing for 4 hours.

d. The painter bills the Project Manager for an extra 4 hours as agreed in the contract.

e. The total amount paid to the painter is $30 which is: $10 (for actual work done painting) + $20 (paid time during weather delays).

f. Although the Project Manager paid $30 for the work, the Earned Value of the work is only $10 because that is the value of work performed.

g. In EVM, we call the cost variance (CV). CV is calculated by subtracting Actual cost from Earned Value which means $CV = EV - AC$

h. $CV = \$10 - \$30 = -\$20$ *(note there is a negative sign!)*

i. In addition to this variance of cost, there is a schedule variance. The work is late!

"This is a very simplistic example but this is how cost overruns occur on projects. The unexpected happens!" They all nodded reflectively.

"Sometimes, variances may not always be due to things beyond our control such as force-majeure." Said Brea.

"Absolutely correct. Variances may also be as a result of any of the following." I proceeded to write out again common causes of variances:

- Bad planning
- Leaving the organization vulnerable to risks and cost overruns
- Poorly designed contracts
- Unrealistic cost estimates
- Unrealistic time estimates
- Changes in technology
- Changes in regulations
- Inexperienced project manager
- Laissez-faire project management
- Poor risk management

- Other causes

I decided to end the session with a recap of some of the core definitions of EVM in the handouts I handed to them.

Definition of Earned Value Management (EVM)

Earned Value Management (EVM) is a project management technique used to measure project performance. EVM is used to compare the amount of work *actually performed* to the amount of work *planned to be completed* by a specific period. There are different software systems (both proprietary and off-the-shelf) used to implement Earned Value Management and manage the work being performed. Any of such systems is called an Earned Value Management System (EVMS).

A few key points on EVM:

- EVM measures the value of work performed in terms of the budget planned for the work (i.e. the baseline budget) and how much was actually spent on the work performed. In other words, EVM helps determine what you got for the amount *(money or time)* you spent.

- EVM integrates cost and schedule measurements to derive the project performance metrics.
- EVM uses the following in calculating project performance:
  - Earned Value (EV)
  - Planned Value (PV)
  - Actual Cost (AC)
  - Budget at Completion (BAC)
  - Schedule Variance (SV)
  - Cost Variance (CV)
  - Schedule Performance Index (SPI)
  - Cost Performance Index (CPI)

  EVM metrics used for forecasting future project outcomes include:
  - Estimate to Complete (ETC)
  - Estimate at Completion (EAC)

o Variance at Completion (VAC)

o To-Complete Performance Index (TCPI)

Throughout this book we will cover all of the above metrics and certain methods and rules used to do so.

Team Discussion

1. What are some common causes of variances?
2. Discuss the underlying philosophy of EVM
3. Discuss steps to take if there are project variances in cost or schedule.
4. Which tools or techniques do you think you would need to have in place to have an up to the day analysis of project performance? Describe to your team member how you would proceed in implementing a system that would track performance daily vs. monthly.

"It is always wise to look ahead, but difficult to look further than you can see."

Winston Churchill

Chapter 3: The Way Forward

Since Cindy became a PMP®, SPI on my chores must always be 1

We met a couple of weeks later as we continued to plod through EVM. "Where are you on your projects today?" I inquired. Some could not answer the question. So let me ask you the same question I asked them! Do you know for a fact where you are? What is the percent complete on your projects? If you are a project manager working on large projects, complex projects, projects that challenge your project management abilities with so many activities being run by virtual, cross-functional teams in different time-zones, on different project components and varying technologies, it becomes very important for you to have an accurate, predictable, and consistent approach to giving status on your project.

One of the things I liked about what we did at a major Aerospace firm I worked for, every Tuesday, was meeting with Senior Engineering Executives and thoroughly reviewing our program of projects. As a result of the in-house software we had, we were able to very efficiently look through a program of projects, and through certain key graphs and charts, we were able to see projects that were doing really well, projects that were on track, projects that were over budget and projects that were questionably under budget.

The need for that accuracy and the need to be able to consistently predict with accuracy what was happening on projects were so important, and that is what our meetings thrived on - Being able to predict the future, being able to see the past, and being able to see what was currently happening. In other words, we could see trends on projects, programs or portfolios.

Looking at various reports that we hear from the industry of project management, we can see that not all projects are successful. Several projects fail. In fact, more projects fail than succeed. In my book, *Project Management Mid-Level to C-Level*, I talk about two failed projects, the Virtual Case File Project and the Sydney Opera House Project. I also review a 2009 report from the Standish Group called the Chaos Report. The Chaos Report indicates that several IT projects fail (68%) compared to those that succeed, and the key reason for failure is poor project management practices. Earned value management will take your practices up a notch if it is done the right way by incorporating solid schedules, milestones, scope and budgeting.

The big question project managers are asked is "Where are you on the project?" or "How far do we have to go?" Some managers ask how much has been spent but the pressing question typically is status. Where are you? How do you know where you are on a project? Earned value management affords practitioners a great approach to knowing where they are.

Three big questions often asked by stakeholders:

1. How much money have you spent and what was the baseline cost for this time period?
2. What percentage of the work have you completed within this time period?
3. Are you on schedule or behind?

So we're talking about *cost*, percent of agreed *scope of work complete* and the *schedule* aspect. That is the traditional triple constraint! Time, cost, and scope. The really great thing about Earned Value Management is that it enables you to predict with accuracy considering these three factors, time, cost and scope.

You see, if you try to give status based on any of those components in isolation, your conclusions would lack validity and conclusive information. The additional information lacking can easily be gleaned through a thorough earned value analysis.

Earned value analysis will enable you to predict with accuracy based on:

- Time (where you are in the schedule)
- Cost (how much money you've spent in relation to time and scope complete)
- Percentage of agreed scope of work complete (in relation to cost and time).

Earned value analysis doesn't give you an analysis in isolation of time cost, or scope. It gives you an analysis based on time, cost, and scope together as *one unit*.

One thing I quickly realized as I got immersed in EVM is that it is very powerful and useful in the right hands. In order for EVM to work, estimates of time, cost and scope must be as accurate as possible. If not, it becomes a case of garbage in garbage out. If one does not plan thoroughly on the front-end, EVM could be the death of a PM's reputation. He or she would forever be described as a lousy planner until they prove themselves not to be! The more accurate your estimates are, the more accurate and meaningful your earned value analysis will be.

Team Discussion

1. Explain how incorrectly calculated cost and schedule can affect an earned value analysis?

2. Explain how incorrect scope estimates can affect earned value analysis.

3. What are some of the ways middle & senior management can take the leadership role in an Earned Value Management implementation?

"On any project, there is no such thing as too much project management."

Phill C. Akinwale

Chapter 4: EVM as a Standard

39 more EVM reports to go! I guess it's fast-food again tonight!

E arned value management has become an industry standard to measure how a project is progressing and how to forecast the project's final cost, the project's completion date and also variances on the project such as cost variances and schedule variances. By integrating time, cost, and percent of the scope complete, earned value provides a very consistent and reliable way to evaluate project performance.

Question

Which of the following is more important, knowing how much money you've spent on a particular piece of work, knowing percent complete to date or knowing how well you are performing on the schedule?

Answer

Rephrasing the question, which element of the triple constraint is most important? Is it time, is it cost, or is it scope?

If you think hard enough, you'll realize that you cannot give one answer. And if you do, it's wrong because Earned Value Management integrates all three components of time, cost and scope. EVM compares how much work you said you would accomplish with how much work you have actually accomplished within a particular timeframe. EVM gives you metrics that indicate if the project is performing as planned relative to cost and schedule or not. Remember, when the term earned value is mentioned, it is in reference to the value of work that has been completed.

Question

On a two day project, I told senior management I would accomplish $5K worth of work the first day. The first day, I only accomplished $3K worth of work, which means $2K worth of work was undone. What is the Earned Value for Day 1?

Answer

The earned value (EV) for Day 1 is $3. It's as simple as that! There are some other metrics that I'll be introducing later on. But first understand this. Earned value refers to the value of work that has been completed.

EVM metrics provide you with a huge red flag warning. A warning sign to you that your project is going off the cliff.

On a particular program I was assigned to, managing program controls (EVM metrics for several projects), I remember management asking me; "How come we've spent so little? Why have we spent so little money on this project?" And I really couldn't answer the question immediately. I came up with all sorts of possible ideas about why we hadn't spent that much money and it all looked good until I began to dig into the reasons why we had spent so little money.

So, how were we able to find out that we had spent so little money? Senior management was able to find out because the earned value metrics looked "fishy". Very fishy! If I remember correctly, our CPI was well above 1.5 but the SPI was very low. Now, later on, I'll explain what these metrics mean, but for now, I'm trying to illustrate that earned value metrics are a great indicator. If you know what they mean, you'll be able to work wonders with them.

When I realized that CPI was so high and SPI was pretty dismal, I began to dig based on management's directive. And what did I find? I found out that one of the main engineers assigned to this work package in the field, had been charging to the wrong cost center. I also realized that some of the engineers hadn't even been charging their time for the period in which they worked. So they would go weeks without charging their time or logging their time in the system. I got that corrected quite quickly, before it cascaded into even more problems that we couldn't track. Eventually, all the charges hit the right control account and all was well but note; the only way we found out those errors was by EVM.

So these metrics are extremely important. Knowing these metrics will give you the opportunity to take corrective action as needed EARLY! This is why meeting weekly was so essential. Consistent earned value analysis will give you ample time to avoid failure by seeing what's coming down the pike. It is just absolutely invaluable to the project manager to be able to see both downstream and upstream and predict with accuracy the future of the project.

Now, apart from those reasons, a very important reason why earned value is needed when working with government agencies and why it's so big is the *OMB Circular A-11 Part 7 Act*. OMB stands for Office of Management and Budget. You could actually Google this acronym "OMB Circular A-11 Part 7".

The long and the short of this circular is that U.S. government agencies, must use a performance based acquisition management system based on ANSI or the EIA Standard 748 to measure the achievement of cost, schedule, and performance goals. This amounts to EVM and that is why it has become increasingly important. This is something that you cannot compromise if you're working for or with a government agency as a contractor or a project manager. It's only a matter of time before you begin to feel the heat in such situations!

Team Discussion

1. By month 8, you completed $45M dollars worth of work and you spent $50M. What is the Earned Value?

2. You are the CEO of U Inc. Your team says 50% has been spent but only 40% has been accomplished. Explain why and why not you could be concerned or not.

3. What is contained in the OMB Circular A-11 Part 7 Act?

4. You are on track on your project schedule but you are over budget. Which variance is likely to be less: schedule variance or cost variance?

"If you keep on managing projects without monitoring the triple constraint and ever-changing project conditions, you'll wind up in a black hole."

Phill C. Akinwale

Chapter 5: Earned Value Metrics

Let me at him! Let me at him! I want him off the project, Noww!

By the sixth week of our classes, Brea and the team were becoming increasingly curious. The suspense of the 3 cryptic metrics and derivative metrics was keeping them on the edge of their seats in every session.

"So when do we get to learn the REAL stuff Teach?" She asked with a silly grin.

"When you have mastered the arcane arts of the EVM philosophy." I replied back.

"Yeah" The others chimed in. "We are being sucked more into the yearly budget and we would like to get done before the holidays kick in." Said Manculo, a program analyst with 2 inch thick glasses who could rattle off 6 digit additions with ease. He was an amazing lad.

"Woah! Patience patience!" I teased. And then I began. "In Earned Value Management, there are three very important metrics upon which several other derivative metrics are based. These metrics have 2 acronyms, one with 2 letters which is more recent and one with 4 letters." I switched on the projector and beamed the PowerPoint presentation on the screen for the team to see the mysteries they had so patiently waited to learn.

Planned Value (PV or BCWS)

The first metric is called planned value, also known as PV (Planned Value) or BCWS (Budgeted Cost of Work Scheduled). BCWS simply put, is the planned cost for the work scheduled across a specific period of time.

To really understand what PV means or what BCWS implies, think about the acronym. *Budgeted Cost of Work Scheduled* and that's why it's called Planned Value because it deals with the "PLAN" to perform work within a *particular period* at a *particular cost*. So we're talking about a period of time, for example, from 1/1/2012 to 1/31/2012.

In order for planned value to make sense, you need to think about a TIME FRAME, and you need to think about COST for a particular portion of the project scope of work within that time frame. So if I say I will get X% of the scope of work done within Y days for Z dollars, then I have my planned value. So, I need to be thinking about how much work I intend to get done within the time frame and how much the work will cost. And that's what makes planned value so relevant. It puts the TIME FRAME (schedule) into perspective. Planned value really becomes a basis for the measurement of the value of work performed. If PV is off, then the whole analysis becomes meaningless.

Exercise 1 (PV)

We planned that we would have completed $7M worth of work in month 1 on Project A. What is our PV for Month 1 of the project?

Answer

Our PV for Month 1 of the project is $7M.

Exercise 2 (PV)

We planned that we would complete $16M worth of work in month 2 in Project A. What is our PV for Month 2 of the project?

Answer

Our PV for month 2 is $16M.

Exercise 3 (Cumulative PV)

What is our cumulative PV at the end of Month 2 on the project?

Answer

Cumulative PV at the end of month 2 is $23M.

Note: Cumulative means from inception to date.

At the end of Month 2 on Project A, adding up values for PV from both Month 1 ($7M) and Month 2 ($16M), our PV to date is $23M.

Actual Cost (AC or ACWP)

The next metric is known as AC (Actual Cost). There also exists another acronym for it, ACWP (Actual Cost of Work Performed), which is the older acronym for actual cost. Nonetheless, it's very important to know these older acronyms. They can be used interchangeably, depending on the teams that you find yourself working with or the Earned Value Management System you are working on. AC (Actual Cost) or ACWP (Actual Cost of Work Performed) is the actual cost of work performed during a specific time period.

AC refers to how much money you spent to complete the work that has been accomplished within a particular time frame. It is an actual cost of money spent which is different from Planned Value (PV) which is a planned cost of work to be accomplished.

Exercise 1 (AC)

Project A has started and we have spent $5M at the end of Month 1 on Project A. What is our AC for the project at the end of Month 1?

Answer

Our AC at the end of Month 1 is $5M.

Exercise 2 (AC)

Project A has started and in Month 2 alone, we have spent $20M. What is our AC for the project at the end of month 2?

Answer

Our AC for Month 2 at the end of Month 2 is $20M.

Exercise 3 (Cumulative AC)

Referring to Exercises 1 and 2; at the end of Month 2 on Project A, what is our cumulative AC (actual costs to date) for the project?

Answer

At the end of Month 2 on Project A, our cumulative AC (actual costs to date) for the project adding up values for AC from both Month 1 ($5M) and Month 2 ($20M) is $25M.

"Easy?" I asked the team.

"So far, so good" said Brea.

"We've talked about two metrics, planned value, also known as BCWS and actual cost, also known as ACWP, actual cost of work performed. Now, let's throw in a third metric and a very important one. *Read very carefully.*" I said ominously as I pointed to the screen.

Earned Value (EV or BCWS)

The final key metric is EV (Earned Value). The older acronym for Earned Value is BCWP (Budgeted Cost of Work Performed). EV which is the budgeted cost of work performed indicates the value of work that you have completed. In order for Earned Value to be calculated, you must know what the Planned Value is because this will give you a basis to value the work performed or complete. And that's why if you do not know your Planned Value and the percentage of the work complete, you cannot compute Earned Value.

Earned Value measures progress because it is related to the percentage of work complete.

Earned Value Formula:

EV = BAC x % Complete	*Use when BAC is given.*
EV = PV x %Complete	*Use when the PV is given. EV should be for the same time period.*

Where:

- *PV = Planned Value*

- *BAC = Budget at Completion*

- *% Complete is the percentage of work complete at the project reporting date.*

Exercise 1 (EV)

We completed $6.5M worth of work by the end of month 1 on Project A. What is EV for the project at the end of month 1?

Answer

EV for the project at the end of month 1 is $6.5M.

Exercise 2 (EV)

We completed $8.5M worth of work in Month 2. What is EV for the project in Month 2 alone (not counting Month 1)?

Answer

EV for the project in Month 2 alone (not counting Month 1) is $8.5M.

Exercise 3 (Cumulative EV)

What is the cumulative EV for the project at the end of month 2?

Answer

At the end of Month 2 on Project A, our cumulative EV (Earned Value to date) for the project adding up values for EV from both Month 1 ($6.5M) and Month 2 ($8.5M) is $15M.

In order to better understand this, let's look at an example that includes the 3 metrics; AC, PV and EV

Example 1 (Illustrating PV, AC and EV)

(PV)

On a two day project, assume we were scheduled to accomplish $5 worth of work on Day 1 and $5 on Day 2. The BCWS for Day 1 is $5, and the BCWS for Day 2 is $5. The cumulative PV for days one and two is $10.

*Side note, we call this cumulative PV for the whole project, the Budget at Completion (BAC).

(AC)

On our two day project, on day one we planned to spend $5 to get a specific portion of the work done. This $5 is the PV or BCWS. However, something went wrong, and we ended up spending more than we had planned, as it happens on several projects.

Our actual cost, or AC, was $7. So we have a scenario for day one of the budgeted cost of work scheduled being $5, but the actual cost of work performed, (how much money we actually spent), is $7.

So we planned to spend $5 to accomplish a certain amount of work valued at $5 within that time period. However, we spent more!

Now, this is the big puzzle here; although I planned to spend $5 on day one and I spent $7...how much work did I actually get done? That would be the big question. So, in other words, what is the percent complete?

(EV)

We completed 50% of the work for Day 1. Realize that EV (Earned Value), or BCWP is equal to $2.50 because we got just half of the work scheduled for Day 1 done. The total work scheduled for completion on Day 1 was $5.

So you've got three metrics here,

1. PV: the amount that we planned to spend within the time period on day one, $5.
2. AC: How much money did we actually spend on day one? $7.

3. How much work did we actually get done? $2.50. Note: We got 50% of the work done. The value of this work is 50% of the Planned Value ($5).

So in other words, to calculate EV, you need to know the percentage of work complete, and that is one of the most subjective aspects of earned value. How do you assign percent complete when you're coding or when you're carrying out other tasks that aren't discretely separated or not easily measured? We'll be talking about that later, but for now, understand the three main metrics of Earned Value Management, PV, AC, and EV.

Budget at Completion (BAC)

One final metric to discuss in this section is the Budget at Completion. This is the total budget for the entire project. It can be described as the total budgeted cost of work scheduled. Adding up PV or

BCWS across the duration of a project will give you the BAC.

Example 2

The total budget for the entire project is $100M. The budgeted cost of work scheduled for each month is divided as follows:

Engineering Project R12	PV or BCWS
January	$7M
February	$23M
March	$30M
April	$15M
May	$15M
June	$10M
Total Budgeted Cost (BAC)	$100M

Formula: *BAC = Total PV or BAC = Total BCWS.*

For example, BAC for the 6 month project starting in January is: PV (Jan) + PV (Feb) + PV (March) + PV (April) + PV (May) + PV (June)

*Remember: PV is also referred to as BCWS.

BAC can be used to calculate the Earned Value on a project from inception to date. In the table given previously for Engineering Project R12, BAC = $100M, assume 40% of the total project work is complete at the end of March, then Cumulative EV = 40% x $100M = $40M.

Uses of Cumulative Earned Value Data

Several organizations practicing EVM may routinely collect, analyze and review data for PV, EV and AC over time either cumulatively or monthly.

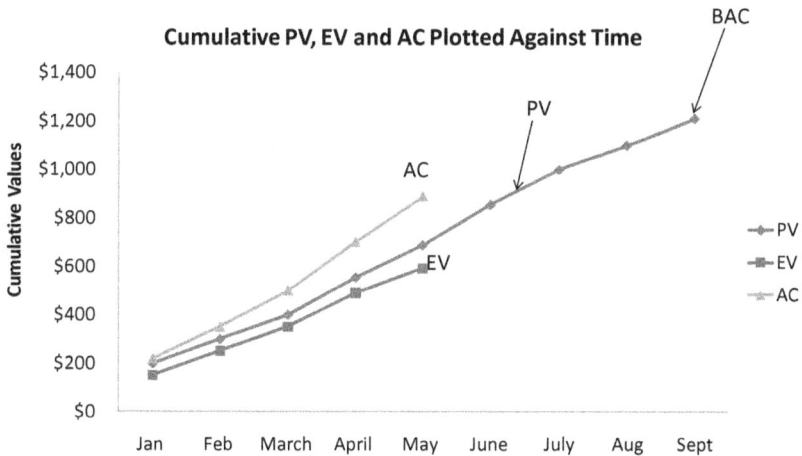

Cumulative PV, EV and AC Plotted Against Time

Team Discussion

1. Briefly define and discuss:

   a. PV

   b. AC

   c. EV

2. Planned Value for the time period is $50 and the percent complete is 50%. What is the Earned Value?

3. In Month 1, Earned Value is $40 and the percent complete is 25% of work complete for Month 1. What is the Planned Value for the time period?

4. On a project, Earned Value is $400 and the percent complete is 30% of work for the entire project. What is the project's BAC?

"Efficiency is doing things right; effectiveness is doing the right things."

Peter F. Drucker

Chapter 6: The Final Verdict: SPI, CPI, CV and SV

Come here you sick bug! Enough changes already!

A t the end of the hour, the team was dazed. I had given them the fire hose to drink from but now they were buzzing and ready to roll again. I continued the lecture:

"Earned Value considers the cost budgeted to perform the work, not the actual cost spent. In other words, however much you end up spending has no bearing on the earned Value."

Brea and the others nodded in agreement.

"Now, the values we've computed on our 2 day project are all well and good. On day one, the metrics are as follows" I said pointing at the whiteboard as I proceeded to write:

- PV = $5
- AC = $7
- EV = $2.50

I continued, "The big question however, is how we use these metrics? In order for these metrics to make sense, we must compute some derivative metrics. These derivative metrics (variances and indices) will reveal if your performance is either good or bad. So let's take a look." And with that, I switched the projector on and they gazed up.

EVM VARIANCES

CV (Cost Variance)

The first derivative metric is cost variance, CV. Cost Variance (CV) is the numeric variance between

Earned Value (EV) and the Actual Cost (AC) for a specific time period.

In other words, CV is the difference between the Earned Value (EV) and the Actual Cost (AC) for a specific time period. CV measures project cost performance.

Formula: $CV = EV - AC$ *or* $CV = BCWP - ACWP$

Remember: EV is also referred to as BCWP and AC is also referred to as ACWP

If CV is:	*This means:*
Positive	Cost is less than budget: work is under budget
Negative	Cost is more than budget: work is over budget
Zero	Cost is equal to budget: work is on budget

If Cost Variance is negative, that means you're over budget. If the variance is positive, that means you're under budget. And if you've got a zero variance, it means you're right on budget.

SV (Schedule Variance)

Schedule Variance (SV) is the numeric variance between the Earned Value (EV) and Planned Value (PV) for a specific time period. In other words, it is the difference between the EV and PV for a specific time period. SV measures project schedule performance.

Formula

$SV = EV - PV$ or $SV = BCWP - BCWS$

Note: EV is also referred to as BCWP and PV is also referred to as BCWS

If SV is:	This means:
Positive	Work is ahead of schedule
Negative	Work is behind schedule
Zero	Work is on schedule

"How was that?" I asked.

"Just groovy" said Cindy. The others nodded.

"Great! Let's talk about some additional derivative metrics. These metrics are different from what we previously discussed. Previously we talked about variances. Now we're talking about indices. And the next slide popped up:

EVM INDICES

Schedule Performance Index (SPI)

The next metric is Schedule Performance Index (SPI). SPI is the ratio of the Budgeted Cost of Work Performed (BCWP) to Budgeted Cost of Work Scheduled (BCWS) for a specific time period. In other words, SPI is the ratio of the Earned Value (EV) to the Planned Value (PV) for a specific time period. SPI measures schedule efficiency on a project.

Formula: $SPI = EV \div PV$ or $SPI = BCWP \div BCWS$

Remember: EV is also referred to as BCWP and PV is also referred to as BCWS

If SPI is:	This means:
Greater than 1	Project is ahead of schedule
Less than 1	Project is behind schedule
Equal to 1	Project is on target

Cost Performance Index (CPI)

Cost Performance Index (CPI) is the ratio of Budgeted Cost of Work Performed (BCWP) to Actual Cost of Work Performed (ACWP) for a specific time period. In other words, CPI is the ratio of the Earned Value (EV) to the actual cost (AC) for a specific time period. CPI measures cost efficiency on a project.

Formula: CPI = EV ÷ AC or CPI = BCWP ÷ ACWP

*Note: EV is also referred to as BCWP and AC is also referred to as ACWP

If CPI is:	This means:
Greater than 1	Cost is lower than budget
Less than 1	Cost is higher than budget
Equal to 1	Project is on target

Remember in my example earlier on real-world EVM, I talked about CPI being greater than 1.5? Well, that meant my project was under budget, seriously under budget, and that was of great concern to senior management. So, depending on the project and circumstances, CPI being greater than 1 could be either good or bad, depending on the situation.

Summary of Formulas

Did you notice that EV comes first in all equations for variances and indices?

- When calculating a variance, you subtract.
- When calculating an index, you divide.
- If the word "schedule" is involved, PV is present
- If the word "cost" is involved, AC is present.

On the next page is a cheat sheet for these 4 formulas.

SPI	EV/PV
CPI	EV/AC
SV	EV – PV
CV	EV - AC

- EARNED VALUE IS KING
- THEY ALL START WITH EV
- IF IT'S A VARIANCE USE SUBTRACTION
- IF ITS AN INDEX, USE DIVISION
- IF IT STARTS WITH AN S, THINK S FOR SCHEDULE (PLANNED VALUE)
- IF IT STARTS WITH A C, THINK C FOR COST, WHICH IS ACTUAL COST

EV

Exercise 1

You are the Project Manager on a project.

At month 2: PV = $10 Million, AC = $8 Million, EV = $4 Million

- What are the SPI and CPI in this scenario?
- What are the SV and CV?
- What should be done to correct the current SV and CV?

Answer

SPI = EV/PV

SPI = $4M/$10M = 0.4

SPI = 0.4

The team is only 40% efficient

CPI = EV/AC

CPI = $4M/$8M = 0.5

CPI = 0.5

The team is only achieving $0.50 worth of work for each dollar spent on the project. The team is not performing well.

SV = EV - PV

SV = $4M – $10M = -$6M

SV = -$6 Million

In monetary terms the schedule is $6 Million behind. If this additional 6 million dollars worth of work had being completed to date, the team would not be behind schedule.

CV = EV - AC

CV = $4M - $8M = -$4 Million

CV = -$4 Million

The team is $4 million dollars over budget.

Strategy for unsatisfactory performance:

- Find causes of variance.
- Get the causes under control and get the schedule back on track.

- Have regular project reviews.

- Think of options such as crashing and fast tracking as a way of catching up on lost time if possible and if applicable.

Exercise 2

You are the PM on a 2 day project.

At day 2: PV = $5 Million, AC = $2 Million, 50% of the work has been done.

- What are the SPI and CPI in this scenario?

- What are the SV and CV?

Answer

Earned Value (EV) = %complete x PV for the time period

PV is the budgeted cost of work scheduled

If 50% of the work is done and 5 million dollars worth of work was scheduled for completion at this time, it means:

EV = 50% x $5M

EV = 0.5 x $5M

EV = $2.5 Million

SPI = EV/PV = 2.5/5

SPI = 0.5

The team is only 50% efficient schedule wise. They are slow compared to the original plan. They may have under-estimated the duration needed to complete the work or they have other problems.

CPI = EV/AC = 2.5/2

CPI = 1.25

The team is achieving 1.25 dollars worth of work for every dollar spent.

SV = EV – PV = $2.5M -$5M

SV = -$2.5 Million

The team is $2.5 Million behind schedule in monetary terms.

CV = EV – AC = $2.5M - $2M

CV = $0.5 Million

Even though the work is $0.5 Million under budget the team is behind schedule.

I switched off the projector much to sighs of relief. "Now you understand 4 derivative metrics of the 3 core Earned Value Metrics." I said. "The toughest part is almost over. Or is it?"

Team Discussion

1. Write out formulas for the following:
   - CPI
   - SPI
   - CV
   - SV
   - EV

2. BCWS is also known as?

3. BCWP is also known as?

4. Define BAC

5. SPI = 0.9 and CPI = 0.9, PV=500. What is AC?

"Unless commitment is made, there are only promises and hopes; but no plans."

Peter F. Drucker

Chapter 7: Benefits of EVM

Imagine a project team without EVM!

It was week 10 of our EVM sessions and the team had already started applying EVM to existing projects. "Are you starting to see the benefits of EVM?" I asked them.

"Well not in executing but certainly in planning out my scope and schedule for the project." Said Manculo with excitement. "I am more aware of the promises I make and if I can truly fulfill them not only at cost but during the specified time period."

"Good point Manny!" I said, dimming the overhead lights in preparation for the presentation. The projector beamed the presentation:

Depending on the nature and size of projects you may find yourself on, the benefits of earned value are centered on the following points.

1. *Early warning signs of cost, schedule, or scope issues.* Remember the project I told you about in which CPI was above 1.5? The only reason management was able to see that something was amiss and the only reason why that issue became visible (we realized that the team was charging to a wrong cost center) was because we were using earned value for project performance tracking and reporting. Without earned value metrics, there's no way management would have realized soon enough that some engineers were charging to a wrong cost center.

2. *Early warning signs of internal and external issues.* On various projects where different teams touch project information are and are involved on the project, there could be interface issues slowing down the project schedule or contributing to project failure as far as the timely completion of project deliverables is concerned. Earned value management makes you ask questions about why work is not proceeding as planned and why things are going wrong, either from an internal or external perspective.

3. *Earned value management provides accessible data to facilitate effective decision making on the project and on the business.* For example, if earned value has been persistently low on a project, CPI and SPI have flat lined at 0.6 or 0.5 over a period of a month, it's a red light prompting teams to stop work and ask questions such as: Will performance get better? Why is performance persistently bad? If an answer that would remedy

the situation is not forthcoming, it could be a sign to shut down the project. If the project cannot recover from that state of low schedule efficiency and cost overruns, it could be a sign to reassess and reevaluate the project. It could also be a sign that the project was incorrectly baselined from the get go, meaning that the schedule, cost (baseline budget), and the scope of work to be complete were not in harmony. It could also be a sign that the project manager and the team may have misunderstood some elements of the work, and the team members who provided the estimate may have also misunderstood the work. In summary, earned value metrics give you a double check on where you are, especially if you are progressing behind schedule and over budget and getting less done than planned within a particular time period.

4. *EVM improves stakeholder confidence by being more transparent and accountable to stakeholders*

on project progress. In instances where EVM is practiced correctly, it becomes very easy for stakeholders to quickly assess project progress, likely future performance, and to assess if present conditions can be remedied in some way. In the grand scheme of things, team communication, team integration, and understanding project performance improves through thorough EVM and thorough Earned Value Analysis (EVA). EVM includes the whole picture of managing projects with Earned Value Metrics but EVA is specifically the end to end analysis that is performed using derived metrics cumulatively and over specific performance periods. EVA is a silver bullet that assists the project team in preventing or reacting to project performance issues. With EVM, you can be sure that if your team is not meeting expectations in time, cost, or rate of work, very quickly those facts will be brought to the surface.

Team Discussion

1) Discuss with your team members in broad terms the benefits of EVM with regards to:

   a. Warning signs

   b. Stakeholder comfort and confidence

   c. Decision making

2) Explain the benefits of observing trends in EVM indices and variances through a thorough EVA.

"Setting PV is about making promises, your resulting EV and AC is all about keeping or breaking them."

Phill C. Akinwale

Chapter 8: Earning Rules

You EVM illiterate! What do you know? I'm the PM!

Earning rules is a critical component of earned value that should be addressed early in the EVM process. A major task in EVM is deciding how to split the work into discreet components, and in order to do this, the team should have a solid Work Breakdown Structure (WBS) and schedule. The WBS is broken down into control accounts, and under each control account, you have various WBS components. A well - planned and defined Control Account is another key aspect that makes EVM work sink or swim.

In a WBS that has been correctly planned and created, the control accounts enable the project manager and the team to very quickly see the health of each control account by looking at its SPI or CPI. Of course, under each control account, you have other WBS components and scheduled work packages that roll up into the control account, but in previous meetings I've been in, I have discovered that senior management is first of all interested in the health of the project and then depending on the project health, they proceed to investigate each control account meticulously.

We had control account managers for each control account, and the health of each control account was the responsibility of the control account manager. So think about it. The WBS is hierarchical.

It looks like a family tree, so imagine the level under the root note being the control account. And under each branch leading up to the control account, you have several other components. In the illustration below, A Class, B Class and C Class are control accounts.

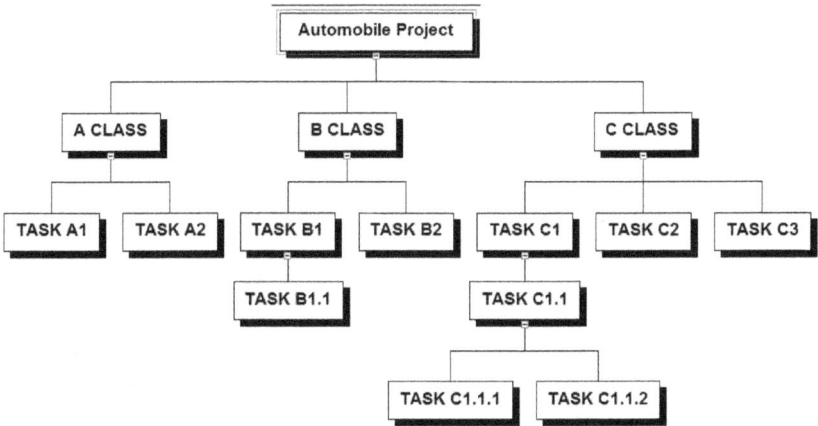

In order to efficiently implement earned value management, you need to know when you can claim a particular piece of work as complete. For example:

- How do you know on an Automobile project, with a hardware design work package, when the hardware design is 50% complete?

- How do you know the percent complete to assign to a software coding work package? Perhaps there are thousands of lines of code with varying degrees of difficulty. How do you know when to claim 10% complete or 0%?

There are various ways you can do this, and that is what we will be studying in this section of earned value management.

Identifying earning intervals up-front reduces the level and the amount of work to complete during the earned value process, and that is why this is a very important step that every team or every project manager should consider. Let's examine the different methods.

1. Weighted Step or Weighted Milestone Method

The weighted step method works best where work to be performed in a work package can be broken down into specific milestones. These milestones serve as check-in points. These check-in points enable the team to claim the right amount of earned value for the milestones that have been completed.

Example

The team is given a work package to perform the following:

1. Conduct a feasibility study for the engineering department
2. Write a report
3. Respond to comments on the study from the team
4. Submit a final document to management for approval
5. Get the final document approved by management.

Let's say for each milestone, the team claims 20% of the budget as earned value, which means for each of those milestones, a particular percentage of the work is deemed complete. So when you add up all those amounts, you get 100% of the work done in the work package.

For example,

ID	TASK	WEIGHT
1	Conduct a feasibility study	20%
2	Write a report	20%
3	Respond to comments	20%
4	Submit final document	20%
5	Get document approved	20%

Assume the budget is $50,000. On completion of milestone one, earned value is $10,000 on completion of milestone two, earned value is $10,000, and the same for all the other milestones.

So you're looking at a breakdown of five milestones, 20% complete per milestone and an earned value of $10,000 may be claimed any time a milestone is complete.

At completion of milestone one, 20% of work is complete. At milestone two, a cumulative value of 40% of the work is done; at milestone three, a cumulative value of 60% of the work is done. At milestone four, 80% of the work is done and at milestone five, 100% of the work is done.

ID	TASK	Cumulative Weight	Cumulative EV
1	Feasibility study	20%	$10,000
2	Write a report	40%	$10,000
3	Respond to comments	60%	$10,000
4	Submit final document	80%	$10,000
5	Get document approved	100%	$10,000

This particular approach is very good in instances where you can break down the work. It requires you to think and set measurable steps and criteria for the advancement of the work and the claiming of earned value.

Weighted Step Method Pros and Cons

- The good thing about this method is instead of saying, "Well, you're not done, so you're not going to get any earned value accounted for this particular work package!" this method allows for partial credit against the tasks or the sub-tasks within that work package.

- On the downside, it requires a lot of up-front planning, thought, effort and documentation of the criteria to measure against.

2. Remaining Effort Method

The remaining effort approach determines the physical percent complete by calculating it. In order to accurately define the physical percent complete, the project team must focus on quantifying the work remaining in a very accurate way. This method essentially reviews the work remaining which has to be accomplished to complete the particular work package or activity. It could be a very long process depending on the nature of work and work package size.

The work remaining in a particular work package or task when quantified in dollars is known as the Estimate to Complete, ETC. Once the estimate to complete is understood, at that point we can say that the amount of work that has been done is xyz dollars. So instead of thinking about it as *I have completed $10 worth of work,* you're thinking about it like this: *In a $20 budget, I have $12 worth of work to complete.* That means I've actually only completed $8 worth of work or my budget is flawed and I am bound to have a negative Variance at Completion (VAC). As time proceeds, conditions may change and the project could still come in on budget but as a PM, always prepare for the worst and hope for the best. That really is the mindset; looking at the estimate to complete the remaining work as opposed to the percent complete.

Remaining Effort Method Pros and Cons
Example of the remaining effort method
The team is assigned to clear a lot. The lot could be broken down into two pieces, piece one and piece two. Both pieces are exactly the same size. The

budget for the work is $5 million. The plan is as follows:

- On day one, $2.5 million of work is planned to be completed,
- On day two, $2.5 million worth of work is planned to be completed, making the full sum for the project.

On day one, the team gets done with half of the lot, and the team has spent $2.5 million. Everything looks good on the surface. However, when the project manager and the inspector begin to review the work, they discover that the second half of the lot will take $3 million to complete. Apparently there were some obstacles that they hadn't previously identified that will need to be removed. At this point, even though day one's activities were on track, the remaining effort needed for the work is more than what was planned. So, in other words, we cannot truly say we are 50% complete using the Remaining Effort method. As far as the schedule and budget is concerned, it appears we are half-way on the surface but digging deeper, we're actually going to go over the budget although the schedule may remain intact. When you really look at the project from this perspective, it gives more realistic results tailored to the present circumstances.

The advantage of the remaining effort method is that it allows the team to be objective and to have a second take on measuring each activity. It also allows for partial credit against the different activities. The disadvantage of this method is that it requires an assessment for each activity. Also, estimating the percent complete requires the team to perform an analysis of the remaining resources.

3. Level of Effort (LOE) Method

I have often viewed the level of effort method, also known as LOE method as a very easy method to implement. However, it is not the best method to use in all situations. The LOE method is based on the passage of time and used very frequently for activities such as project management or other services. In this method, EV and PV are always equal.

Month	PV	EV
Jan	$150	$150
Feb	$100	$100
March	$75	$75
April	$50	$0
May	$75	$0
June	$55	$0

Report Date →

In the LOE method, EV and PV are always equal.

Example 1(LOE)

An engineer is working on a 100 hour work package. At hour 80, the work package status is given as 80% of the scheduled time having elapsed, and therefore, EV is 80%.

Example 2 (LOE)

On a 30 day work package, the status date is Day 24. The project team writes a report and states that EV is 80% because 24 out of 30 days have elapsed.

LOE Pros and Cons

- As observed, the advantages of this method are it does not require a status check. It is very simple to immediately say the Earned Value is x or y.

- The disadvantage is it provides no real measure of progress and therefore can be questioned by management, especially when earned value always matches planned value, regardless of what was completed or not. It is not the most reliable of earned value methods to use.

4. *Units Produced Method*

Assume a project to produce four widgets. Two widgets are complete. Earned value is 50%. If all four widgets cost $40,000, then the earned value is $20,000 (50% of $40,000).

Units Produced Pros and Cons

- The advantage of this method is it is straight forward. It's objective. It's easy and very transparent.

- The disadvantage of this method is that it is relegated simply to a production type environment where similar items have fixed unit prices.

- Another disadvantage here is that it does not consider labor fluctuation. In one instance, it might take five hours to produce a widget. In another instance, the same widget could take a longer for other reasons.

5. **Fixed Formula Method**

The fixed formula method is used on activities that are very easy to assess and analyze, and typically are short term. It requires minimal effort.

The downside of this effort is it is not effective for long term endeavors. Let's take a look at some variants of this method. Others exist.

i.   *0/100 Rule*

Consider a project in which a certain task needs to be performed. A team may decide that no Earned Value will be claimed until this task is complete.

ii.   *50/50 Rule*

50% is claimed for starting a task and the other 50% is claimed for completing the task.

iii.   *25/75 Rule*

25% is claimed for starting a task and 75% is claimed for completing the task

iv.   20/80 rule

20% is claimed for starting a task and 80% is claimed for completing the task

Fixed Formula Method Pros and Cons

- The fixed formula method works well in instances where the work is short term. It doesn't require much effort. It's very easy to calculate.

- Where long term items are concerned, it is not as effective as other methods such as the weighted step method.

6. Percent Complete

This method is used in situations where the physical progress of a project task can be measured. The physical progress of the task determines its earned value. It is best used in situations where completion of project work can be easily measured.

Example

The project team is working on a $2,000 work package to clear a lot. Half of the lot is cleared. The Earned Value is $1,000.

Percent Complete Pros & Cons

- Easy to implement. Detailed planning of work-package is not required as assessment is based on the cost account manager's assessment or project manager's assessment.

- On the flipside, assessing percent complete is not always as clear-cut as in the previous example. It could be very subjective due to the lack of detailed planning.

Team Discussion

1. Discuss all Earned Value methods and rules. Which ones do you remember without looking at the book?

2. Which Earned Value Method fits your current projects best?

3. Which approach or approaches do you think you should adopt widely across the organization and why?

"If you have to forecast, forecast often."

Edgar R. Fiedler

Chapter 9: Forecasting

Boy! I can't wait for my vacation! After all I've done my time.

Forecasting with Earned Value metrics enables project managers to see what is coming ahead by looking at the past and present to predict the future. Forecasting involves project personnel carrying out a variance analysis and a trend analysis to better predict future project performance and take corrective action where needed.

Forecasting involves making predictions regarding the project's future in terms of estimates to complete the work or estimating the total amount likely to be spent at project completion based on current performance. Note that this is different from deducing the initial project budget which is not based on the team's actual performance.

EVM FORECASTING METRICS

Estimate at Completion (EAC)

EAC is the total amount of money predicted to be spent at project completion based on current performance. Note EAC is predicted after the project work has commenced. EAC is the actual cost of work that has been completed so far plus the estimate of the cost of the remaining work to be done in order to complete a project or a specific activity in a project phase.

There are several ways to calculate EAC depending on the data available. Let's examine each of these methods.

EAC FORMULAS

i. *Manual bottom-up summation (When past performance shows assumptions were flawed)*

In this case the Project Manager computes a new ETC from ground up.

Formula:

$EAC = AC + ETC$

ii. *EAC (assuming ETC work performed at the budgeted rate)*

$EAC = AC + BAC - EV$

iii. *EAC (assuming ETC work performed at the current CPI)*

$EAC = BAC/Cumulative\ CPI$

iv. *EAC (considering both SPI and CPI)*

$EAC = AC + [(BAC-EV)/(cumulative\ CPI \times cumulative\ SPI)]$

Estimate to Complete (ETC)

Estimate to Complete (ETC) is the expected or estimated cost of the remaining work that needs to be done in order to complete a project or a specific activity in a project phase. There are several ways to calculate ETC depending on the data available.

ETC FORMULAS

ETC = EAC –AC (*Use when only EAC and AC values are available*).

*Bearing in mind that **if work is being performed at the present cumulative CPI**, we could incorporate the following formulas in to the equation:*

EAC = BAC/ CPI

CPI = EV/AC which means: AC = EV/ CPI

If work is being performed at the present cumulative CPI, the formula (ETC = EAC – AC) could then be modified as follows:

ETC = (BAC / CPI) – (EV / CPI)

ETC = (BAC – EV) / CPI

(Use when current variances are typical and work is being performed at the present CPI).

On the other hand if current variances are atypical and ETC work will be performed at the budgeted rate.

$$ETC = BAC - EV$$

Summary of EAC Formulas and When to Use Them
EAC = AC + ETC
Use when: Past performance shows assumptions were flawed. A new bottom-up ETC is used.
EAC = AC + BAC – EV
ETC work will be completed at the budgeted rate. Current variances are not typical and will not continue in the future.
EAC = BAC ÷ Cumulative CPI
Use when: Variances are typical and current cumulative CPI is expected to remain the same in the future
EAC = AC + [(BAC – EV) / (Cumulative CPI x Cumulative SPI)]
Use when: Considering both **cumulative** CPI and SPI factors where schedule may also impact ETC

☀ Scenarios for EAC and ETC

Where: AC = actual cost to date, CPI = cumulative CPI, EV = cumulative earned value to date, BAC = Budget at Completion, SPI = cumulative SPI

EAC scenario 1: Manual Bottom-up Summation for ETC

- Jack realizes his budgets for the data center are flawed at month 6 in a 12 month project.

- It is June and Jack has spent $50,000 (AC) out of a $75,000 (BAC) project.

- Obviously his assumptions are flawed somewhere.

- Jack does a bottom-up estimate to calculate how much the remaining work will cost (ETC).

- He assesses the remaining work (ETC) will cost $60,000.

- He adds this ETC to the amount spent so far. EAC = AC + ETC

- EAC = $50,000 + $60,000 = **$110,000**

- This gives Jack a realistic EAC using a totally re-estimated ETC.

EAC scenario 2: ETC Work Performed at Budgeted Rate Scenario

- Jack realizes his budgets for the data center need to be tweaked.
- It is June and Jack has spent $50,000 (AC) out of a $75,000 (BAC) project.
- Jack assesses the value of work completed so far.
- Jack has spent $50,000 (AC) but he has only accomplished $35,000 (EV) worth of work.
- He is sure the amount of remaining work (ETC) will be completed at the budgeted rate.
- He subtracts value of work done (EV) from total budget (BAC) to calculate remaining work.
- Estimated Cost of Remaining work (ETC) = BAC − EV = $75,000 - $35,000 = $40,000
- EAC = AC + (BAC − EV)

- Jack has spent $50,000 (AC) and has $40,000 (BAC – EV) work left to complete.

- EAC = $50,000 + $40,000 = *$90,000*

- *This gives Jack a realistic EAC only if ETC work will be performed at the budgeted rate.*

EAC scenario 3: ETC Work Performed at Current CPI

- Jack realizes his budgets for the data center need to be tweaked.

- It is June and Jack has spent $50,000 (AC) out of a $75,000 (BAC) project.

- Jack assesses the value of work completed so far.

- Jack has spent $50,000 (AC) but he has only accomplished $35,000 (EV) worth of work.

- The cumulative CPI on this project is CPI = EV/AC = $35,000/$50,000 = 0.7

- If the current CPI continues into the future, then the EAC will be affected

- EAC based on this CPI is EAC = BAC/CPI = $75,000/0.7

- EAC = $107,142.86

- *This gives Jack a realistic EAC only if ETC work will be performed at the current CPI.*

EAC scenario 4: ETC Work Considering SPI and CPI (gives a more realistic estimate)

Jack decides to factor in both SPI and CPI to get a better feel for the ETC. He sees this as a possible worst case scenario especially since SPI and CPI are both less than 1.

- Jack realizes his budgets for the data center need to be tweaked.

- It is June and Jack has spent $50,000 (AC) out of a $75,000 (BAC) project.

- Per the schedule, the amount of work to have been completed by now (PV) is $37,500

- Jack has spent $50,000 (AC) but he has only accomplished $35,000 (EV) worth of work.

- The cumulative CPI on this project is CPI = EV/AC = $35,000/$50,000 = 0.7

- The cumulative SPI on this project is SPI = EV/PV = $35,000/$37,500 = 0.93
- If the current cumulative SPI and CPI continue into the future, then the EAC will be affected
- EAC = AC + [(BAC − EV)/(Cumulative SPI x Cumulative CPI)]
- EAC = $50,000 + [($75,000 - $35,000)/(0.93 x 0.7)]
- EAC = $50,000 + [$40,000/0.651] = $50,000 + 61,443.93
- *EAC = $111,443.93*

Note the different scenarios and how a different EAC was used based on a different set of assumptions. Note the different set of assumptions produced different results when factored in. This illustrates that understanding which formula to use for EAC is key. If you are presented with AC and ETC, then use the formula EAC = AC + ETC, but if presented with other data, read very carefully and deduce the correct formula to use.

Variance at Completion (VAC)

Variance at Completion (VAC) is the numeric variance between the budget at completion (BAC) and the estimate at completion (EAC). It forecasts the likely cost variance will be at the end of the project when all the work is completed.

VAC Formula: $VAC = BAC - EAC$

If VAC is	This means:
Positive	Project is likely to finish under budget
Negative	Project is likely to finish over budget
Zero	Project is likely to finish on budget

To-Complete Performance Index (TCPI)

To-Complete Performance Index (TCPI) is a forecast of cost performance that must be attained on remaining work in order to meet a specified management objective such as the budget at completion (BAC) or estimate at completion (EAC). It represents the level of cost efficiency that is required on the remaining work in order to achieve BAC or EAC. It is the ratio of work remaining to funds remaining.

$$TCPI = \frac{Work\ Remaining}{Funds\ Remaining}$$

- If the BAC is no longer viable, a forecasted EAC is computed.
- If the EAC is approved, it supersedes the BAC as the cost performance goal.
- TCPI = work remaining/funds remaining
- TCPI based on the BAC = (BAC-EV)/(BAC-AC)
- TCPI based on the EAC = (BAC-EV)/(EAC-AC)

TCPI Exercise

1. Calculate the TCPI in the following scenario
2. Advise if the BAC is still viable or not.
3. Compute a new EAC based on an ETC of $7M

BAC = $10M

EV = $5M

AC = $8M

Answer

TCPI (based on BAC) = ($10M - $5M) / ($10M - $8M) = 5/2

TCPI = 2.5

TCPI is 2.5 This means that to meet management objectives, $2.5 worth of work must be accomplished for every dollar spent) which is virtually impossible on a normal project

To deduce the viability of the BAC, the cumulative CPI is needed.

Cumulative CPI = EV/AC = 5/8

CPI = 0.625 (CPI is not used to calculate TCPI but is used to assess viability of BAC and how optimistic or conservative the TCPI is).

- Since the TCPI (2.5) is much greater than the current CPI (0.625) by a large margin. It means future efficiency must improve significantly if the project is to achieve the Budget at Completion (BAC). Comparing cumulative CPI to the TCPI It is apparent that the BAC is no longer viable.

- In that regard, the Project Manager computes a new EAC since the BAC is unrealistic at this point.

Compute a new EAC based on an ETC of $7M

Estimating the work left to be completed at $7M

AC = $8M

ETC = $7M

EAC = AC + ETC = $8M + $7M = $15M

The newly forecasted EAC is $15M

TCPI based on the EAC = (BAC-EV) / (EAC-AC)

TCPI (EAC) = ($10M - $5M) / ($15M - $8M) = 5/7 = 0.71

TCPI = 0.71

This means that to meet management objectives, $0.71 worth of work must be accomplished for every $1.00 spent *which sounds achievable though a little optimistic compared to the current low 0.625 cumulative CPI.*

Performance Reviews

Performance Reviews are reviews comparing cost performance and schedule performance to their respective baselines. This includes variance analysis (what caused the cost or schedule variance and how can it be corrected), trend analysis (what is the performance trend over time and is project performance improving or getting worse? Performance reviews consider earned value performance (taking into consideration all relevant earned value metrics).

Project Management Software

This is used to track the key earned value management metrics (PV, EV and AC) and associated indices and variances which include SPI, CPI, CV and SV.

EVM metrics used for forecasting future project outcomes include:			
Estimate to Complete (ETC)	Estimate at Completion (EAC)	Variance at Completion (VAC)	To-Complete Performance Index (TCPI)

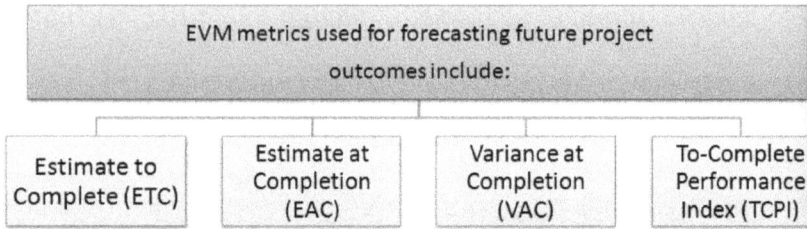

Summary of forecasting metrics

Team Discussion

1. What are the advantages of forecasting using EVM metrics?

2. Describe how you would use cumulative values of SPI, CPI, CV and SV to predict project performance and take corrective or preventive action.

3. What is TCPI and what is its relevance? In other words, what does TCPI indicate?

4. Write out the formulas for:

    a. EAC (all 4 formulas)

    b. ETC

    c. VAC

    d. TCPI

"All successful people have a goal. No one can get anywhere unless he knows where he wants to go and what he wants to be or do."

Norman Vincent Peale

Chapter 10: Understanding Trends

Bubba took looking for trends a bit too far…

One of the biggest things any project manager could do with Earned Value Metrics is to use them to observe trends from the past and present and predict how the future will be. In order to do this, we will plot graphs with Earned Value metrics on the Y axis and time on the X axis. There could be various types of data used for EVA with graphs. Let's start off with the typical representation of PV, EV and AC over time.

Example 1

Plot a graph of the following metrics for Project R.5.12 over 6 months. What can be concluded about this project in general?

Month	PV	EV	AC
Jan	$4,000	$3,000	$5,000
Feb	$5,000	$5,500	$6,000
March	$6,000	$5,750	$6,500
April	$7,000	$7,300	$7,000
May	$8,000	$7,525	$7,725
June	$10,000	$10,000	$10,000

The image below is a cumulative plan for performance (PV)

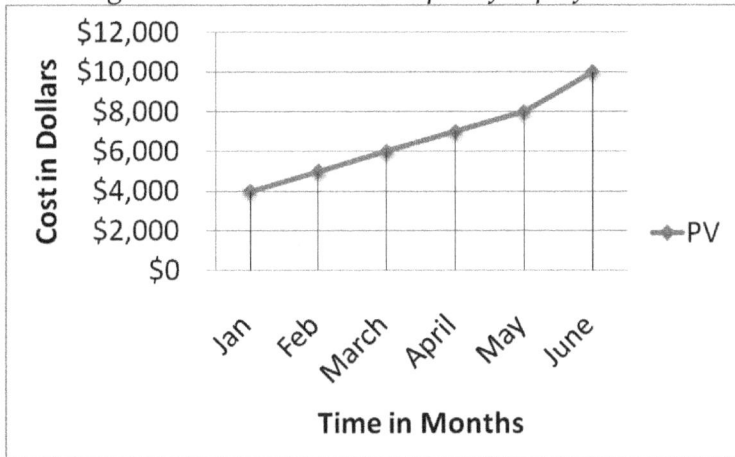

The image below represents actual results: Cumulative PV Compared with Cumulative AC and EV

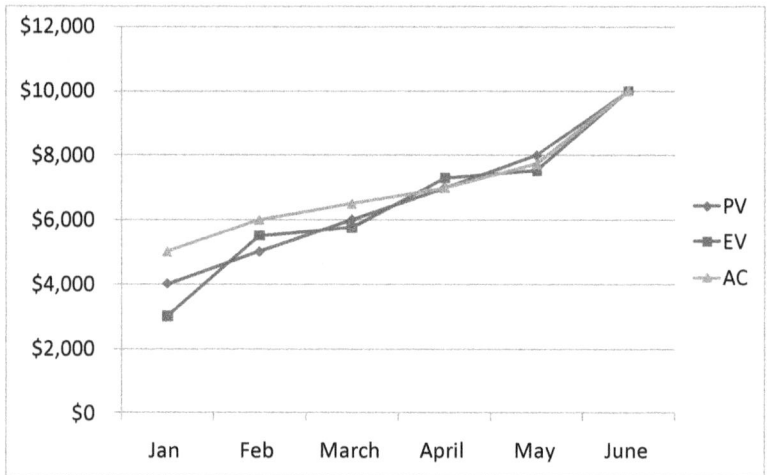

See the following page for a larger graph of the same Project R.5.12.

- In which months did the team perform ahead of schedule?
- In which months did the team perform behind schedule?
- In which months did the team perform over budget?
- In which months did the team perform over budget?

Earned Value Analysis of Project R.5.12

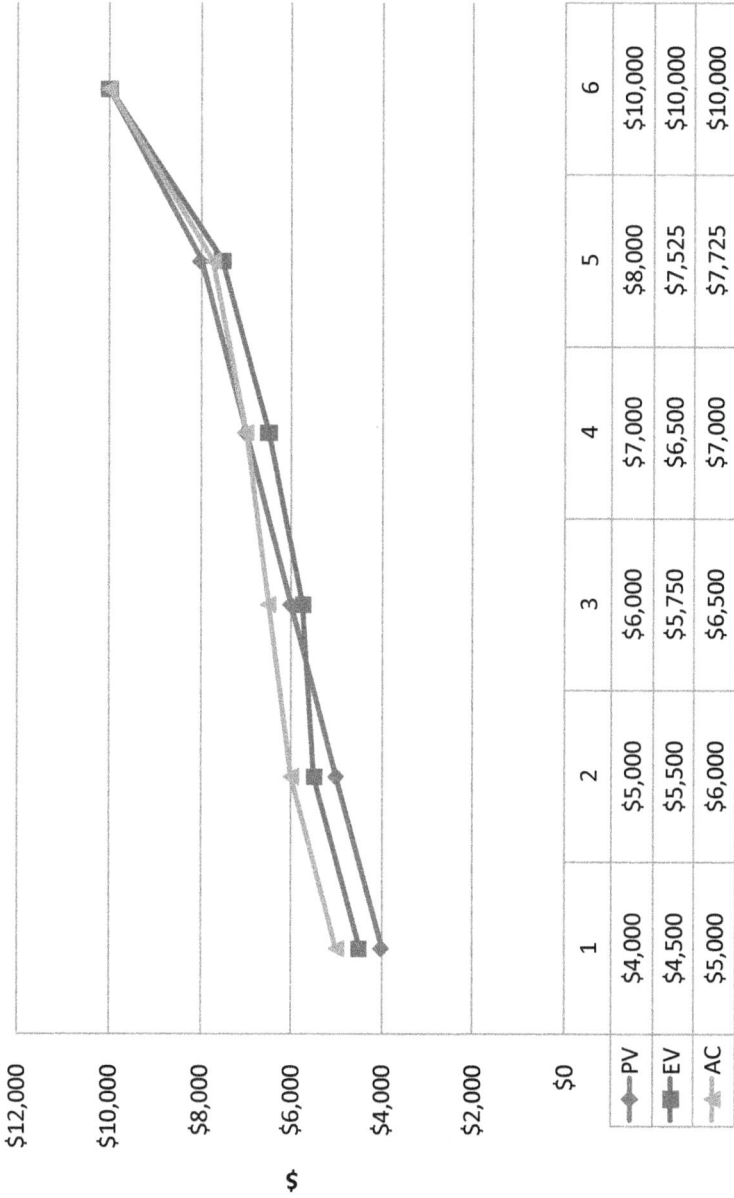

	1	2	3	4	5	6
PV	$4,000	$5,000	$6,000	$7,000	$8,000	$10,000
EV	$4,500	$5,500	$5,750	$6,500	$7,525	$10,000
AC	$5,000	$6,000	$6,500	$7,000	$7,725	$10,000

119

Now let's take a look at the same project but this time with a plot of CPI and SPI over time.

Plot of CPI over time

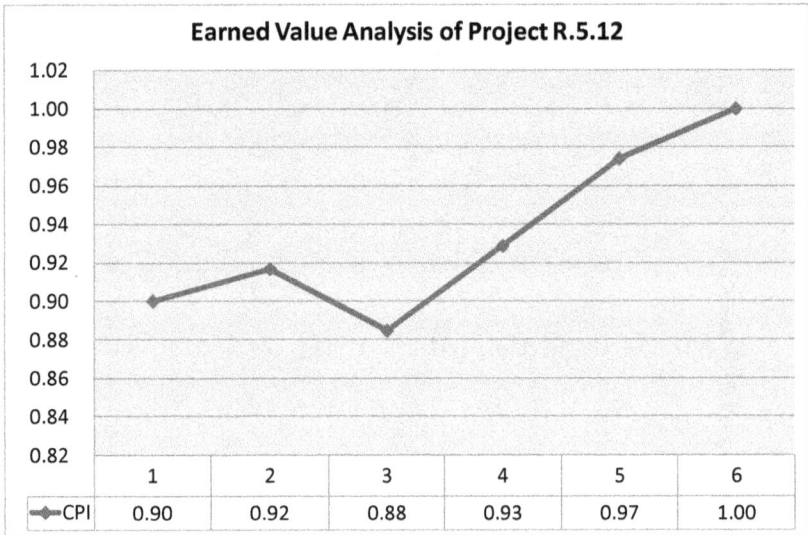

Earned Value Analysis of Project R.5.12

	1	2	3	4	5	6
◆ CPI	0.90	0.92	0.88	0.93	0.97	1.00

Looking at the graph for project R.5.12 of CPI over time, we see that the team under-performed all the way through until the end. There was a dramatic increase in cost efficiency after month 3 following a significant dip after month 2. Management could ask questions such as, why did the team's efficiency drop so drastically? Why was the team never once on budget? Once the reasons are clear, the team could be directed to take further preventive actions to ensure such dips do not recur on the project.

Plot of SPI over time

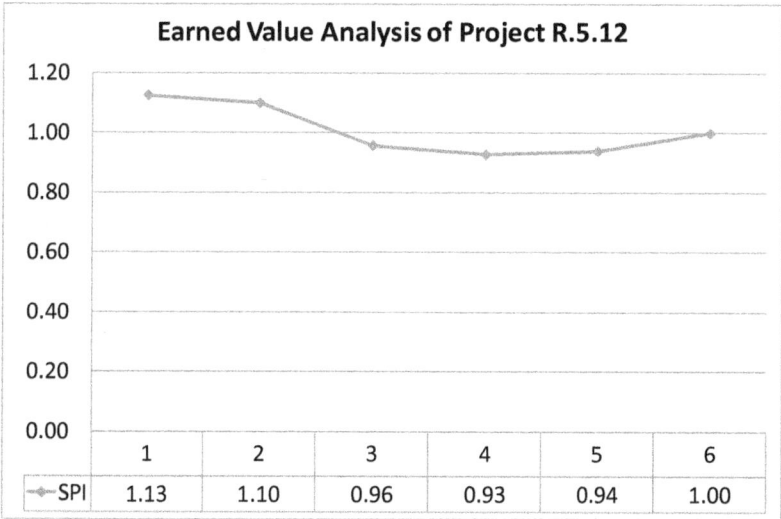

Earned Value Analysis of Project R.5.12

	1	2	3	4	5	6
SPI	1.13	1.10	0.96	0.93	0.94	1.00

Looking at the graph, it can be observed that after month 2, the team under-performed through the rest of the project although they did better in the final month.

In the first month, SPI was rather high and in subsequent months it began to drop considerably and then picked up again at the end of the project. In a typical project, a graph like this would have management asking questions from the get-go such as; why is our SPI so high? Why did the team's efficiency drop so drastically even after month 2 when it could have stayed normal at 1?

Several lessons can be learned by asking the team the right questions, probing for more information and digging for details at every level of the WBS.

Team Exercise

1. What is the value of work performed in June?

Month	PV	EV	AC
Jan	$4,000	$3,000	$5,000
Feb	$5,000	$5,500	$6,000
March	$6,000	$5,750	$6,500
April	$7,000	$7,300	$7,000
May	$8,000	$7,525	$7,725
June	$10,000	$10,000	$10,000

2. From the graph for project A.13.5, calculate values for the following in the first 3 months:

a) CPI

b) SPI

c) CV

d) SV

e) BAC

3. Discuss what is likely to happen in May and June in the following EVM analysis scenario.

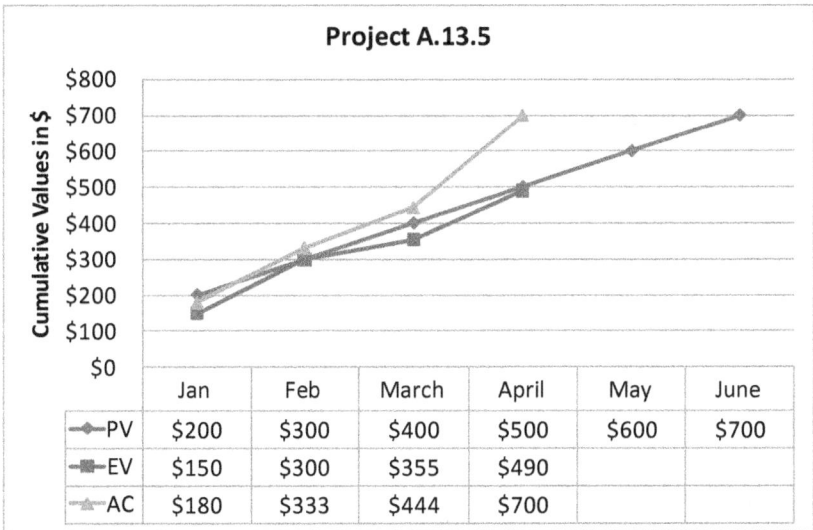

Project A.13.5

	Jan	Feb	March	April	May	June
PV	$200	$300	$400	$500	$600	$700
EV	$150	$300	$355	$490		
AC	$180	$333	$444	$700		

Cumulative Values in $

4. In May and June, interpret what happened from the graph for Project A.13.5 on the next page.

Project A.13.5

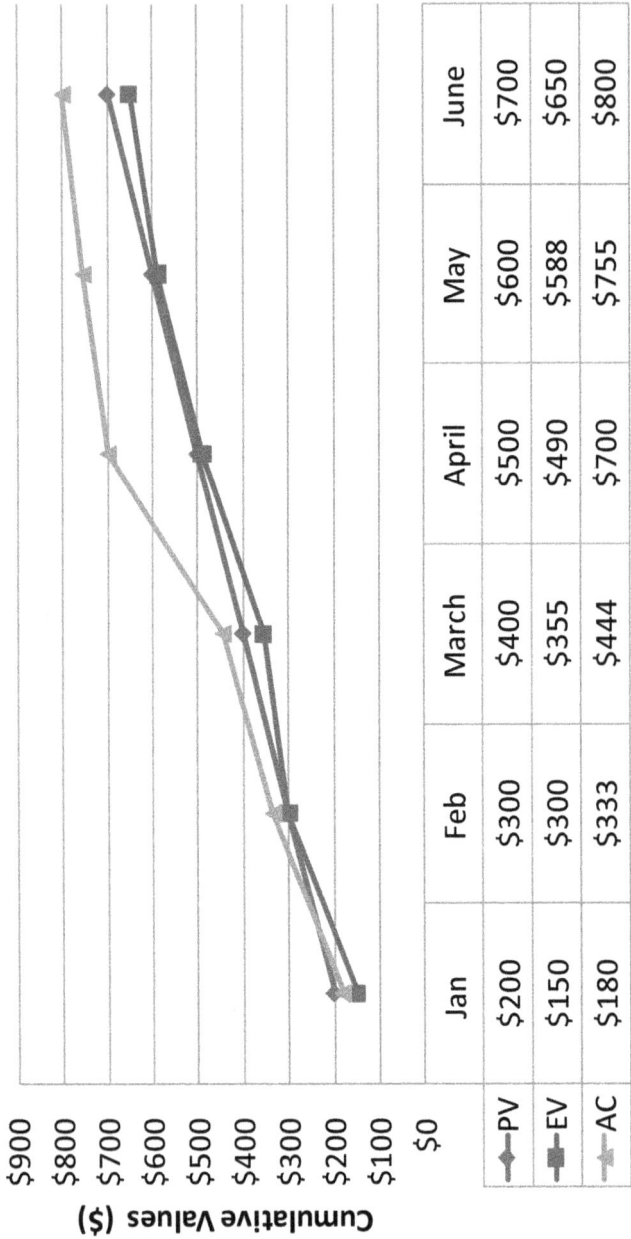

	Jan	Feb	March	April	May	June
PV	$200	$300	$400	$500	$600	$700
EV	$150	$300	$355	$490	$588	$650
AC	$180	$333	$444	$700	$755	$800

Team Discussion

1. Give some examples of graphs you can plot with EVM metrics on both the X and Y axis.

2. What should management do during EVM meetings?

3. Plot graphs of 3 project management scenarios at any point in time where the project is:

   a. Under budget

   b. Over budget

   c. On budget

   d. Mark the point as "P".

"Flaming enthusiasm, backed up by horse sense and persistence, is the quality that most frequently makes for success."

Dale Carnegie

Chapter 11: Conclusion & Next Steps

Take the bull by the horns and stay on it!

E arned Value Management, though abstract at first

is not so difficult that it cannot be implemented on

some simple small-sized projects right away.

Contrary to perception, EVM doesn't need

millions of dollars to implement. All it needs is a solid

understanding of the process as well as a reliable procedure for:

1) *Breaking down the work in the WBS into control accounts and work packages in a meaningful well thought-out way.*

2) *Scoping out the work over the project's duration with meaning and deliberateness.*

3) *Accurate activity duration estimating for each work package.*

4) *Accurate assessments of physical work complete and work remaining.*

5) *Applying appropriate earned value rules. A work package could apply just one or all earned value rules.*

6) *Changing the mindset and the culture to understand and embrace EVM.*

7) *Training for the team to truly understand the philosophy. The formulas are easy to grasp but the philosophy is key. True understanding of which makes it easy to manage projects with the EVM mindset.*

8) *Deciding upon the reporting frequency (based on real values available). For some organizations with huge projects, this may be daily, weekly or monthly.*
9) *Decide on which EVMS to use based on your needs and relevant standards.*

Next Steps

The best way to learn the most valuable lessons in earned value is to practice it head-on, hands-on. Without doing so your perception will remain only theoretical, so do the following:

a) Begin to practice it in a scaled down way on your projects and gradually.

b) Begin to apply more from the EVM tool-kit as relevant or as needed.

c) Attend any earned value management classes you can to learn more

d) Look for an Earned Value Management group on LinkedIn or with the PMI®

e) Keep on demonstrating value with Earned Value!

f) Deliver on your project promises using EVM to keep you on the straight and narrow and you, your management and customers will be very happy!

Epilogue

At the end of the final EVM lecture, the team decided to meet bi-monthly to review the latest best practices in EVM. Brea went on to become a PMP® Certified project manager, Manculo became the VP of IT Operations. Cindy became a CAPM® and then a PMP®. Kirk went on to start his own software company in Texas. And the other team members? They are doing just fine. Still plugging away at EVM! As for me I moved on to other endeavors and have had a great time learning more about EVM as the years go by.

Stay focused, keep learning about Earned Value Management and don't forget to put it to practice on your projects!

References

1. The *PMBOK® Guide* Fourth Edition. Published by The Project Management Institute (PMI®) 2008

2. OMB Circular No. A–11 (2008). Published by Executive Office of the President (Office of Management and Budget) June 2008

3. The Practice Standard for Earned Value Management Second Edition. Published by The Project Management Institute (PMI®) 2011.

4. Earned Value Management Guidelines. Washington State Department of Transportation. 2008

5. Project Management Essentials. Published by Praizion Media 2011.

Acronyms

AC – Actual Cost

ACWP – Actual Cost of Work Performed

BAC – Budget at Completion

BCWP – Budgeted Cost of Work Performed

BCWS – Budgeted Cost of Work Scheduled

CPI – Cost Performance Index

CV – Cost Variance

EAC – Estimate at Completion

ETC – Estimate to Complete

EV – Earned Value

EVMS – Earned Value Management System

OMB - Office of Management and Budget

PV – Planned Value

SPI – Schedule Performance Index

SV – Schedule Variance

TCPI – To Complete Performance Index

VAC – Variance at Completion

WBS – Work Breakdown Structure

Also available from Praizion Media

Sketches of a **PMO**

a story of a silly project management office

DVD available at www.praizion.com

Praizion media
Real world project management training solutions

www.ingramcontent.com/pod-product-compliance
Lightning Source LLC
Chambersburg PA
CBHW021238090426
42740CB00006B/584

9 7 8 1 9 3 4 5 7 9 4 0 4